1.0 INTRODUCTION

The process of urbanization is increasing and population growth continues at a rapid pace. Mobility is a fundamental requirement of all human activity whether it falls into either of the two categories of work or play. According to the latest forecast, there will be more than 9 billion people living on our planet in 2050 and approx. 1.2 billion cars will be using the world's roads. The automobile industry's yearly growth rate is expected to exceed 5.5% from 2010 to 2015, reaching a value of more than $5,132 billion by 2015, according to recently conducted research survey. In line with this, the automobile industry has developed into one of the most important business sectors during the past hundred years. Automotive textiles is the one of the most dynamic and promising sector of the industrial textiles. It is the single largest consumer of the industrial textiles with over one million tons per annum. The proportion of automotive textiles in the total market for industrial textiles accounts for approximately 22%. Automotive textiles are used for filters for air and fuels, vehicle awnings, airbag cushions, seat belts, composites for structural components, moulded parts for interiors, interior decoration for moulded parts, seat or protective covers, moulded parts for seats, car floor coverings, drive belts and hoses.

1.1 AUTOMOBILE INDUSTRY – GLOBAL SCENARIO

In 1982, the number of vehicles (cars, trucks and buses) manufactured around the world were 36.2 million. But that figure had risen to more than double as 77.6 million in the year 2010. China was dominating the automobile industry in 2010, producing 18.3 million vehicles, and it has been able to increase its production sixty-fold since the early 1980. China was followed by Japan with 9.6 million vehicles, the USA with 7.7 million and Germany with 5.9 million. According to estimates provided by the Roland Berger management consultancy company, approx. 18 million Chinese will purchase a new car in 2015.

Global market for light vehicle production experienced a drastic fall, the highest hit in the last 30 years during 2008 and even deeper in 2009. Production of cars was 67.4 million in 2008, and fell to 58.6 million in 2009. But China and India saw a profitable market with production increasing by 50% and 17% respectively. Fluctuations in vehicle manufacturing affected the automotive textile market consequently.

1.1.1 AUTOMOTIVE TEXTILES – MARKET SCENARIO

Automobile industry is expanding in a big way. Subsequently, the share of automotive fabrics is also rising. Within this context, automotive textiles can make a very valuable contribution, because they primarily combine the positive properties of lightweight items: good sound insulation, UV resistance, rigidity, formability and resistance to wear. As a result, experts are predicting that textile composites will have excellent growth prospects in vehicle production around the globe. The enormous variety of different ways of using automotive textiles in vehicle manufacturing provides huge economic potential. If the proportion of textiles in a medium-sized car still only accounts for 25 kilograms at the moment, this figure will increase to 30-35 kilograms in just a few years. Proportionately, 50-60 percent will be non-woven fabrics and felts, while 40-50 percent will be other textile fabrics. The use of natural fibres or fibre mixtures in car manufacturing will also continue to grow partly because it is easier to recycle them, but mainly because natural fibres are lighter than chemical fibres and this can lead to weight savings of up to 40 percent per car (Source: IVGT, 2010).

Climate control can be made more efficient by using spacer textiles, particularly in vehicle interiors. New, recyclable composite materials can be used to provide improved sound insulation

– and they are made of non-woven materials, flocked surfaces or membranes. Generally, the current trend with automotive textiles is moving in the direction of safety and functionality. Vehicles need to become safer and safer and at the same time provide innovative functions – e.g. using smart textiles (conductive textiles). About 70 percent of all technical innovations depend on the material properties – i.e. the need for new materials is spurring on research and development.

Manufacturers are currently standardizing their operations to customize the vehicles to satisfy legal requirements, as well as the preferences of their end users. Supplemented with modern technological advancements and a touch of eco concern, the industry will get added push in the forthcoming years, doubling its results, marking an exemplar change. On an average, 165 tonnes of fabrics are used in car manufacturing process every year. The demand for comfort, concern for safety while driving, and an increasing focus towards reducing fuel consumption and CO_2 emissions have all augmented the usage of special textiles in automobiles, especially cars. In a midsized car, 20 kg of textiles were used generally. With the increasing demand for sophistication, and a stringent want for using protective textiles, the use of textile materials for the same has currently become 26 kg, and is even predicted to increase by 35 kg by the end of 2020. Two thirds of the automotive textiles are used in trims, seat covers, roof, door liners, and carpets. Other fabric applications include tyres, hoses, safety belts, and air bags.

Airbags, trims, and truck covers accounts for 28%, proving to be the largest share of the coated fabric demand. US demand for coated fabrics is anticipated to grow 2.1% on a year on year basis, and reach 655 million sq. yards by 2012. Sales for the same will rise 3.4% annually reaching $3 billion by 2012. Increased sales in this sector will stem from the automobile textile sector. Gains in the sale of automobiles will motivate the automotive textile market as well. While non-rubber coated fabrics will also have a good market, rubber coated fabric materials will be in high growth trajectory. Rubber coated fabrics used in automobiles are expected to post enormous gains by 2012. Knitted and woven fabrics will have a dominating share in the global automobile fabric arena, closely followed by composites, and non-woven. Circular knitted fabrics are used in the interiors of cars for seat covers, door panels, headliners, headrests, boot covers, sunroofs, and parcel shelves. Having the virtues of high flexibility, comfort while traveling, stretchability, and high-grade visual quality; these fabrics will have good potential. Woven fabrics will see a profitable market in the making of door panels, seat covers, side door paneling, and headrests.

Lighter cars prove to be more economical, consumer less fuel, and also involve less manufacturing costs. As the concept benefits the manufacturer, consumers, and the environment application of natural fibre composites will have a rapid growth in the automobile fabric industry. New product innovations in auto fabrics sector will further augment its growth. The latest is the creative combination of seatbelts, and airbags. The belt cum air bag is made with round and smooth edges to give added comfort to its wearer. This process uses high volume production of cold gas inflator, making the seatbelt cum air bag to swell almost five times larger than the width of the normal seat belt. The swelling process is initiated through a deployment signal from a sensor system. This technique will minimize the extent of damage, as it distributes the force across the wearers body. The interior fabrics of the car are made with a special kind of micro fibre which is ultra light, flame proof, and is also resistant to abrasion. This is believed to be 100% eco friendly as well.

1.2 MAJOR COMPONENTS OF AUTOMOTIVE TEXTILES

1.2.1 SEAT BELTS

According to World Health Organization (WHO), road accidents are a worldwide problem and now results in over 1.2 million people die every year from road crashes all around the globe. Occupant restraints, is the best solution such as seat belts and air bags, are highly effective in preventing death and injury from traffic collisions. Seat belts are easy to use, effective and inexpensive means of protection in an accident. Seatbelts are expected to reduce the overall risk for serious injuries in crashes by 60-70% and the risk for fatalities by about 45 percent.

Seat belt plays a crucial role in the safety of passenger during sudden accident. The wearing of both front and back seat belts is now compulsory in many countries of the world and all new cars made, contain atleast four diagonal and lap seat belts each made from about 250 g of woven fabric. Due to ever increasing concern on safety of passenger during accidents, seat belts show better holding position in industrial textiles market.

1.2.1.1 Seat Belt –Market Scenario

According to a Market Research firm, the market for all types of narrow fabrics is estimated at US $ 2.4 billion in 2005. This is projected to grow at about 3.6% per annum during 2005 to 2010. A new study carried out in six European cities shows that advanced seat belt reminders can increase seat belt use among drivers in urban areas up to rates of 93% to 100%.

1.2.1.2 Seat Belt – A Life Saving Guard

The seat belt is an energy-absorbing device that is designed to keep the load imposed on a victim's body during a crash down to survivable limits. Fundamentally, it is designed to deliver non-recoverable extension to reduce the deceleration forces, which the body encounters in a crash. Seat belts function as a safety harnesses which secure the passengers in a vehicle against harmful movements during collision or similar incidents.

1.2.1.3 Classification of Seat belts:

Use of belt depends upon weight of passenger, as per passenger's weight belt width are specified by British standards.

BS STANDARD	Application	Shoulder	Lap
BS 3254 Part 1 1988	"Restraining devices for adults" width [mm]	Min.35	Min.46
BS 3254 Part 2 1991	"Restraining devices for children" width [mm]	Weight 9-18 kg	Min.25
		Weight 18-36 kg	Min.38

Table.1 – Classification of Seat belt

1.2.1.4 Seat Belt Dynamics

(a) Work-Energy principle

The change in the kinetic energy of an object is equal to the net work done on the object. For a straight-line collision, the net work done is equal to the average force of impact times the distance traveled during the impact.

$$W_{net} = \frac{1}{2} mv_{final}^2 - \frac{1}{2} mv_{initial}^2$$

Average impact force x distance traveled = change in kinetic energy

If a moving object is stopped by a collision, extending the stopping distance will reduce the average impact force.

(b) Forces acting on passenger during accident with and without Seat belt:

According to Newton's first law of motion an object at rest tends to stay at rest and an object in motion tends to stay in motion with the same speed and in the same direction unless acted upon by an unbalanced force. Thus, driver continues in motion, sliding forward along the seat. A driver in motion tends to stay in motion with the same speed and in the same direction, unless acted upon by the unbalanced force of a seat belt. Seat belts are used to provide safety for passengers whose motion is governed by Newton's laws. The seat belt provides the unbalanced force, which brings driver from a state of motion to a state of rest.

The task of the seatbelt is to stop driver with the car so that the stopping distance is probably 4 or 5 times greater than if driver with no seatbelt. A crash which stops the car and driver must take away all its kinetic energy, and the work-energy principle then dictates that a longer stopping distance decreases the impact force.

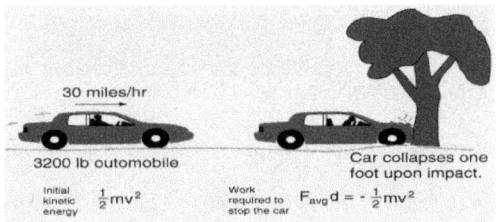

Fig.1- Forces acting during Car Crash

Assume, car speed is 13.41 m/s, driver's weight is 50 kg, then what will be the impact force acting on the car driver?

(i) If driver is wearing a non-stretchable seat belt, stopping distance is 0.304 metres.

Impact force on driver = (0.5*50*13.41*13.41)/0.304 = 14788.5 N = 1.5 tons

Deceleration of driver = 14788.5/50 = 295.77 m/s^2

With no seatbelt to stop the driver with the car, the driver flies free until stopped suddenly by impact on the steering column, windshield, etc. The stopping distance is estimated to be about one fifth of that with a seatbelt, causing the average impact force to be about five times as great. The work done to stop the driver is equal to the average impact force on the driver times the distance traveled in stopping. A crash which stops the car and driver must take away all its kinetic energy, and the work-energy principle then dictates that a shorter stopping distance increases the impact force.

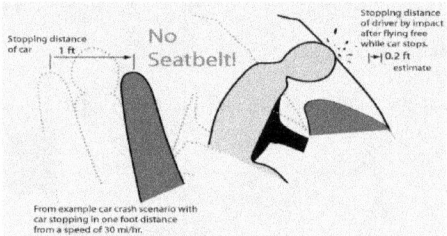

Fig.2 - Force acting without Seat belt

(ii) If driver is without seat belt, suppose stopping distance is 0.0608 metres.

Impact force on driver = (0.5*50*13.41*13.41)/0.0608 =73942.5 N = **7.5** tons

Deceleration of driver = 73942.5/50 =1478.8 m/s^2

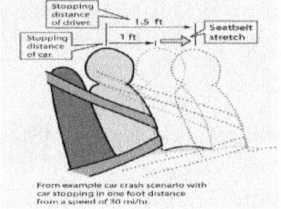

Fig.3 - Forces acting with seat belt

(ii) If driver is wearing stretchable seat belt, stopping distance is 0.456 metres.

Force on driver = (0.5*50*13.41.13.41)/0.456 = 9467.65 N = 0.96 ton

Deceleration of driver = 9467.65/50 = 189.353 m/s^2

A moderate amount of stretch in a seatbelt harness can extend the stopping distance and reduce the average impact force on the driver compared to a non-stretching harness. If the belt is stretched for 0.125metres, it would reduce the deceleration to 189.353 m/s^2 and the average impact force to 0.96 ton compared to 295.77 m/s^2 and 1.5 tons for a non-stretching seat belt. Either a stretching or non-stretching seat belt reduces the impact force compared to no seat belt.

1.2.1.4 Critical Characteristics of Seat Belt:

The critical characteristics of the webbing are abrasion resistance, resistance to light and heat, capable of being removed and put back in place easily and good retraction behaviour. The load bearing capacity of seat belt is 1500 kilograms. The surface of the webbing (generally smooth) is of particular significance because its structure and properties decisively influence the retraction behaviour.

1.2.1.5 Seat Belt – Fibres and Fabric structure

Raw material used for seat belt webbings are Nylon, PET and its derivatives like HTPY. Nylon was utilized in some early seat belts, but due to of its higher UV degradation resistance; polyester is now widely used worldwide. Polyester scores over nylon because of its lower extensibility and higher stiffness.

Seat belt structures can be single layer or double layer and manufactured on needle looms with plain weave, twill weave, satin weave. Normally 2/2-twill structure is used. The needle loom used presently is shuttleless and capable of delivering over 1000 picks per minute.

Material	Body warp yarn	Selvedge warp yarn	Weft yarn	Binder thread	Locking thread
PA/PET	1260 D/108(fil.)	250 D/24(fil.)	500D/48(fil.)	100 D/18(fil.)	250 D/24(fil.)
PA/PET	846/2 D	846 D (mono)	423 D (multi)	220 D	150 D
			396D (mono)		

Table.2 – Seat belt – Fabric Structure

The seat belts are made from nylon filament yarn or polyester filament yarn which is woven to produce the webbing pattern. The linear density of synthetic yarns should be between 100dtex and 3000dtex, preferably 550-1800dtex. The filament linear density should be between 5dtex and 30dtex, preferably 8-20dtex. A typical seat belt is made of 320 ends of 1,100 dtex polyester each. Most weft yarns made from polyester are 550dtex. Commonly used seat belt webbing is a narrow fabric measuring 46mm wide. This structure permit highest yarn packing within a given area for highest strength and sometimes-coarser yarns are used for good abrasion resistance.

1.2.1.6 Manufacturing method:

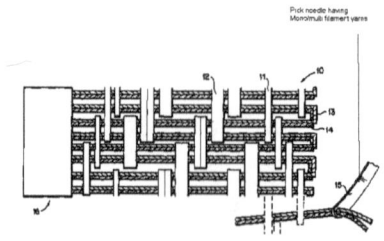 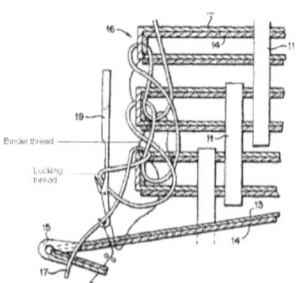

Fig.4 -Pick needle insertion in shed   **Fig.5- Selvedge formation with catch cord yarns**

In the needle loom, the weft is inserted from one side of the warp sheet and here a selvedge is formed. The other side of webbing is held by an auxiliary needle, which manipulates a binder and a lock thread. Once these are combined with the weft yarn, a run-proof selvedge is created. Special care is taken when constructing the selvedge to ensure it is abrasion resistant. It is equally important to ensure that the selvedge is soft and comfortable to wear.

In a loom, needle is used to move monofilament and multifilament weft yarn across width of the web in single shed. A high tension applied to monofilament compared to multifilament yarn so that it doesn't protrude beyond selvedge. Monofilament weft yarns are rigid yarn that provides good stiffness across web and high resiliency. Monofilament weft yarns are under high tension which controls webbing softness and round shape together with catch cord yarns and finer warp yarns in selvedge than body of the webbing. Low longitudinal stiffness and thinner web gives good winding performance with use of smooth monofilament weft yarns. Selvedge warps are having about 66% lower thermal shrinkage than body warp threads. Weft threads are having 16% higher dry heat shrinkage than selvedge yarns. Thus woven ribbon is heated upon after dyeing. This gives soft, round shape selvedge.

1.2.1.7 Finishing:

The woven seat belt webbing is, then transferred under tension to a dyeing and finishing range. The Grey webbing is dyed and heat set whilst webbing made from spun-dyed yarn is heat-set only. Heat-setting is undertaken to impart precise extension to the webbing and suppress its recovery in the event of crash. This is achieved by shrinking the webbing in a controlled manner which increases the weight of webbing from 50 g/m to approximately 60 g/m. Seat belt webbing can be coloured by two methods: either by incorporating spun-dyed yarns or by piece dyeing. The appropriate technique for piece dyeing polyester webbing is by thermosol process.

1.2.1.8 Quality Requirements of Seat Belt:

The seat belt is required to have the following properties:

❖ Static load bearing capacity upto 1500 kg and extensibility upto 25-30%

❖ Abrasion resistance

❖ Heat and light resistance

❖ Light weight

❖ Flexibility for use

1.2.1.9 Performance tests and standards for seat belt webbing:

Tests are carried out for ascertaining the mechanical performance of the webbing as per BSI and the SAE.

- Width
- Thickness
- Breaking strength and elongation for dry and wet webbing
- Abrasion resistance at various environmental conditions and in contact with buckles and fittings
- Influence of extreme environmental conditions and temperatures
- Rubbing fastness
- Microbial resistance

1.2.1.10 Various Defects in Seat belts:

Seat belts are designed to withstand tremendous loads, but if there is any damage to a seat belt its load-bearing ability is significantly reduced.

1. Fraying:

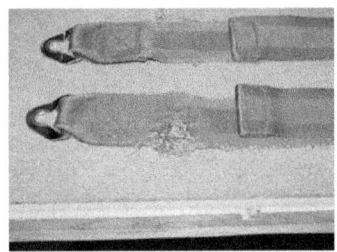

Fig.6 – Seat belt - Fraying

Fraying is one of the most noticeable defects in seat belts. Fraying is where fibres break in the weaving. This results in reduced strength. Seat belts should be repaired or replaced at any sign of fraying. Another form of fraying that is encountered appears as a fuzzy layer on the seatbelt surface. This fuzziness may appear harmless, but it is indicative that fibres are breaking in the weave. The seat belt should be replaced or repaired.

2. UV Damage:

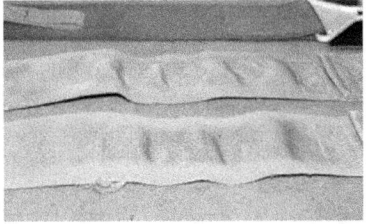

Fig.7 –Seat Belt –UV Damage

Over time, UV rays will damage seat belts. UV damage is more frequent in helicopters and aircraft.

1.2.1.11 Inflatable belts

Ford® introduces the auto industry's first-ever production inflatable seat belts, which are designed to provide additional protection for rear seat occupants, often children and older passengers who can be more vulnerable to head, chest and neck injuries. The inflatable seat belt is an amalgamation of the air bag and the seat belt which is held by weak stitches that burst open when the belt is inflated. Under impact the belt gives 450% more surface area than the normal flat belt. This inflatable rear seat belts spread crash forces over five times more area of the body than conventional seat belts; this helps reduce pressure on the chest and helps control head and neck motion for rear seat passengers These belts could be fitted in the rear seats of the automobile to replace use of air bags in that compartment.

A folded envelop of 3" is covered by 2" face fabric, which is stitched. A folded envelope is rubber coated. Today's envelopes are made from urethane or silicon to make thin, gas leakage proof, high strength, heat resistance and when crash occurs a gas fills and inflates the envelope. The face fabric (web) is made of a resin such as PET. Today's cover fabric uses warp knitted structure which has good strength, good stretching characteristics, comfortable. Warp knit uses yarn of 3000 denier or below. By using such belts kinetic energy of passenger is distributed over a larger area, therefore load experienced by a passenger is small and passenger protected very effectively.

1.2.2 AIR BAGS

An airbag is an occupant restraint system consisting of a flexible envelope designed to inflate rapidly during an automobile collision. Airbags are gas-inflated cushions built into the steering wheel, dashboard, door, roof, or seat of the car that use a crash sensor to trigger a rapid expansion to protect you from the impact of an accident. Airbags are a type of automobile safety restraint like seatbelts. There are several kinds of air bags – driver bag, front passenger bag, thorax bag, curtain bag, rear bag, and knee bag.

In most countries air bags are mandatory for all passenger cars due to stringent in legislation. According to a recent survey, air bags system has contributed up to 20% reduction in fertilities resulting from front collision. Earlier air bags were considered as substitute to seat belts and were only limited up to high speed sports cars. But today, air bags are working in coordination with seat belts. In fact, air bags cushion an occupant in an event of crash. This helps in avoiding the heat on collision.

In 1981, Mercedes-Benz® introduced the airbag in Germany as an option on its high-end S-Class (W126). In the Mercedes system, the sensors would automatically pre-tension the seat belts to reduce occupant's motion on impact (now a common feature), and then deploy the airbag on impact. This integrated the seat belts and airbag into a restraint system, rather than the airbag being considered an alternative to the seat belt. In 1987, the Porsche® 944 turbo became the first car in the world to have driver and passenger airbags as standard equipment.

In Europe, airbags were almost entyrely absent from family cars until the early 1990s. The first European Ford to feature an airbag was the face lifted Escort MK5b in 1992; within a year, the entyre Ford range had at least one airbag as standard. By the mid 1990s, European market leaders such as Vauxhall/Opel, Rover, Peugeot, Renault and Fiat had included airbags as at least optional equipment across their model ranges. During the 2000s side airbags were commonplace on even budget cars, such as the smaller-engined versions of the Ford Fiesta and Peugeot 206, and curtain airbags were also becoming regular features on mass market cars.

1.2.2.1 Air bag – Market Scenario

In 2010 the world market for automotive airbags stood at 258 million units but by 2017 this figure is predicted to rise to 446 million units, representing an increase of 73.2%. Front airbags are expected to remain the biggest category of airbag in 2017. However, forecasts for the seven years to 2017 suggest that they will form the slowest growing category whereas sales of side-impact airbags, inflatable curtains and knee airbags will grow considerably faster. As a result, the share of front airbags in the total market for airbags will drop from 42.2% to 37.9% between 2010 and 2017 although sales of front airbags are still forecast to increase by 55.2%.The fastest increase, at 539.0%, will be in sales of knee airbags. However, these will continue to represent by far the smallest category with sales of just 33 million units in 2017, representing 7.4% of the market. Sales of side-impact airbags are set to rise by 74.7% to 128 million units, representing a 28.6% share of the market, while sales of curtain airbags will increase by 65.2% and account for a 26.1% share.

1.2.2.2 Airbag – Principle of Operation:

Airbags inflate, or deploy, quickly faster than the blink of an eye. In the first 15 to 20 milliseconds after the accident, airbag sensors detect the crash and then send an electrical signal to fire the airbags. Typically a squib, which is a small explosive device, ignites a propellant, usually sodium azide. The azide burns with tremendous speed, generating nitrogen, which inflates the airbags. Within 45 to 55 milliseconds the airbag is supposed to be fully inflated. Within 75 to 80 milliseconds, the airbag is deflated and the event is over.

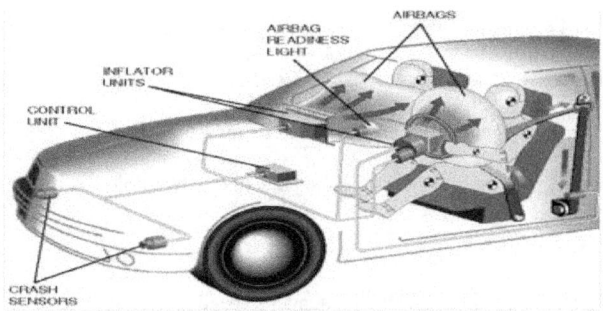

Airbag – Principle of Operation

When airbags work properly, they dramatically reduce the chance of death or serious injury. However, the speed with which airbags inflate generates tremendous forces. Passengers in the way of an improperly designed airbag can be killed or significantly injured. Unnecessary injuries also occur when airbags inflate in relatively minor crashes when they're not needed.

1.2.2.3 Laws of Motion:

The moving objects have momentum (the product of the mass and the velocity of an object). Unless an outside force acts on an object, the object will continue to move at its present speed and direction. Cars consist of several objects, including the vehicle itself, loose objects in the car and, of course, passengers. If these objects are not restrained, they will continue moving at whatever speed the car is traveling at, even if the car is stopped by a collision.

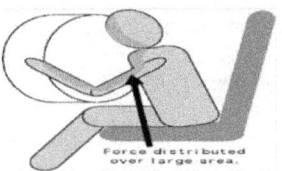

Fig 2: Force distribution on body during collision

Stopping an object's momentum requires force acting over a period of time. When a car crashes, the force required to stop an object is very great because the car's momentum has changed instantly while the passengers' has not there is not much time to work with. The goal of any supplemental restraint system is to help stop the passenger while doing as little damage to him or her as possible.

What an airbag wants to do is to slow the passenger's speed to zero with little or no damage. The constraints that it has to work within are huge. The airbag has the space between the passenger and the steering wheel or dashboard and a fraction of a second to work with. Even that tiny amount of space and time is valuable, however, if the system can slow the passenger evenly rather than forcing an abrupt halt to his or her motion.

1.2.2.4 Airbag Inflation:

The goal of an airbag is to slow the passenger's forward motion as evenly as possible in a fraction of a second. There are three parts to an airbag that help to accomplish this feat:

- The bag itself is made of a thin, nylon fabric, which is folded into the steering wheel or dashboard or, more recently, the seat or door.

- The sensor is the device that tells the bag to inflate. Inflation happens when there is a collision force equal to running into a brick wall at 10 to 15 miles per hour (16 to 24 km per hour). A mechanical switch is flipped when there is a mass shift that closes an electrical contact, telling the sensors that a crash has occurred. The sensors receive information from an accelerometer built into a microchip.

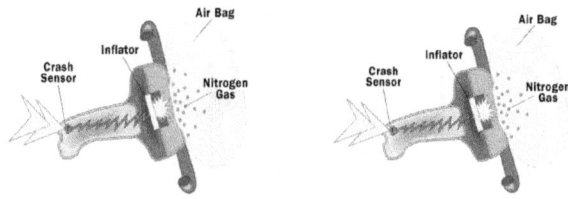

Fig 3: The airbag and inflation system stored in the steering wheel

An air bag contains a mixture of sodium azide (NaN_3), KNO_3, and SiO_2. A typical driver-side airbag contains approximately 50-80 g of NaN_3, with the larger passenger-side airbag containing about 250 g. Within about 40 milliseconds of impact, all these components react in three separate reactions that produce nitrogen gas. The reactions, in order, are as follows.

$$2 \, NaN_3 \rightarrow 2 \, Na + 3 \, N_2 \, (g)$$

$$10 \, Na + 2 \, KNO_3 \rightarrow K_2O + 5 \, Na_2O + N_2 \, (g)$$

$$K_2O + Na_2O + 2 \, SiO_2 \rightarrow K_2O_3Si + Na_2O_3Si \, (silicate \, glass)$$

Hot blasts of the nitrogen inflate the airbag.

Early efforts to adapt the airbag for use in cars bumped up against prohibitive prices and technical hurdles involving the storage and release of compressed gas. Researchers wondered:

- If there was enough room in a car for a gas canister
- Whether the gas would remain contained at high pressure for the life of the car
- How the bag could be made to expand quickly and reliably at a variety of operating temperatures and without emitting an ear-splitting bang

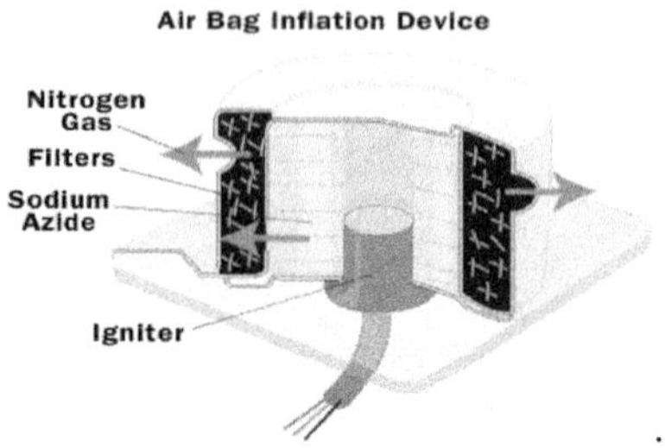

Air Bag Inflation Device

Nitrogen Gas

Filters

Sodium Azide

Igniter

Fig 4: The inflation system uses a solid propellant and an igniter

The airbag system ignites a solid propellant, which burns extremely rapidly to create a large volume of gas to inflate the bag. The bag then literally bursts from its storage site at up to 200 mph (322 kph) faster than the blink of an eye. A second later, the gas quickly dissipates through tiny holes in the bag, thus deflating the bag so you can move.

Even though the whole process happens in only one-twenty-fifth of a second, the additional time is enough to help prevent serious injury. The powdery substance released from the airbag, by the way, is regular cornstarch or talcum powder, which is used by the airbag manufacturers to keep the bags pliable and lubricated while they're in storage.

1.2.2.5 Types of Airbags:

1. Air bags For Frontal Injury:

Air bags are typically designed to deploy in frontal and near-frontal collisions, which are comparable to hitting a solid barrier at approximately 13–23 km/h (8–14 mph). Roughly speaking, a 23 km/h (14 mph) barrier collision is equivalent to striking a parked car of similar size across the full front of each vehicle at about 45 km/h (28 mph). This is because the parked car absorbs some of the energy of the crash, and is pushed by the striking vehicle.

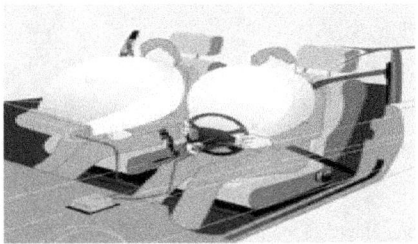

Fig 5: Airbag system for Frontal Collision

Air bags in certain car models deploy twice, for two crashes; it first deploys and deflates, and then re-inflates upon a subsequent collision. The sound of air bag deployment is very loud, in the range of 165-175 dB for 0.1 second. Hearing damage can result in some cases.

2. Side Air bags for cars:

The Side air bags, a relatively new technology designed to protect the head and/or torso (chest and abdomen) in side-impact collisions, are becoming increasingly common in automobiles. Risk was reduced when cars with head/ torso air bags were struck by cars/minivans (significant) or pickup trucks/sport utility vehicles (non significant).

Fig 6: Side Airbags for Cars

Risk was reduced in two-vehicle collisions and among male drivers and drivers aged 16–64 years. Protective effects associated with torso-only air bags were observed in single-vehicle crashes and among male and 16- to 64-year-old drivers.

3. Airbags for Motorcycle:

Honda Motor Co., Ltd announced it has succeeded in developing the world's first production motorcycle airbag system. The Motorcycle Airbag System is comprised of the airbag module, which includes the airbag and the inflator; crash sensors, which monitor acceleration changes; and an ECU, which performs calculations to instantly determine when a collision is occurring.

Fig.7-Airbags in Motorcycle

The airbag module, containing the airbag and inflator, is positioned in front of the rider. The airbag ECU, positioned to the right of the module, analyzes signals from the crash sensors to determine whether or not to inflate the airbag. Four crash sensors attached on both sides of the front fork detect changes in acceleration caused by frontal impacts.

4. Pedestrian Protective Air Bags:

The airbag device for pedestrian protection according to the present invention is able to prevent a pedestrian mounted on the hood panel from falling down on road surface. The airbag device for pedestrian protection according to the present invention is mountable on a motor vehicle with a hood panel, and includes an airbag inflatable in front of the vehicle.

1.2.2.6 Raw materials (fibres) for Air bags:

Air bag fabrics are made of Nylon 6, 6 multifilament yarns with counts from 235 to 940 Tex. The number of monofilaments ranges from 70 to 220. Air bag fabrics are generally dense. Tensile strength, elongation, tear propagation resistance and weight requirement of air bag fabric are critical. Air permeability of air bag fabric should be uniform across the whole width of the fabric. There are currently two principle material types which are used in the manufacture of airbags. They are uncoated nylon (polyamide 66) and coated nylon. Two types of commonly used coatings are Silicone and Neoprene. In general, coated materials are used for driver's side airbags and side impact bags, while passenger side airbags are made from uncoated nylon materials. The advantages of the coated type are better nongas permeation, easier bag-pressure control, and greater heat resistance to burning particles. The noncoated type is lighter in weight, thinner, more flexible, and less expensive.

Table 1: Comparison of Fibre properties of Air bag

Properties	Nylon6,6	Polyester
Density(kg/m^3)	1140	1390
Specific Heat Capacity (kJ/kg/k)	1.67	1.3
Melting Point (°C)	260	258
Softening Point (°C)	220	220
Energy to melt (kJ/Kg)	589	427

Although nylon 6,6 and polyester have similar melting points, the large difference in specific heat capacity causes the amount of energy required to melt polyester to be about 30% less than that required to melt nylon 6,6. Hence in any inflation event that uses a pyrotechnic or pyrotechnic-containing inflator, cushions made from polyester yarn are far more susceptible to burn or melt through in the body of the cushion or at the seam.

The second advantage of nylon 6,6 is its lower density. Lower mass has key advantages – reducing the mass of the cushion lowers the kinetic energy of impact on the occupant in out-of-position situations, thus enhancing safety, while allowing the overall weight of the vehicle to be reduced.

1.2.2.7 Manufacturing of Airbag fabrics:

An airbag for deployment outside of a vehicle is made from a fabric that is coated on at least one of the surfaces of the fabric. The airbag fabric may be coated to increase the strength of the fabric so that the fabric is reinforced. A coated airbag base fabric made of a textile fabric that has an excellent air barrier property, high heat resistance, improved mountability, and compactness and excellent adhesion to a resin film is characterized in that at least one side of the textile fabric is coated with resin, at least part of the single yarns of the fabric are surrounded by the resin, and at least part of the single yarns of the fabric are not surrounded by the resin. An airbag is characterized by using such a coated airbag base fabric. A method for manufacturing the coated airbag base fabric is characterized by applying a resin solution having a viscosity of 5–20 Pas (5000– 20,000 cP) to the textile fabric using a knife coater with a sharp-edged coating knife at the contact pressure between the coating knife and the fabric of 1–15 N/cm.

1.2.3 AUTOMOTIVE FILTERS

The performance of an automobile is bolstered significantly by the presence of high performance filter media in the engine compartment and in various other locations. The purpose of filter is to control contamination, through achieving a balance between the sources of contamination and the ability of a system to tolerate contamination. The ultimate goal is to balance filtration performance with the desired cleanliness level.

1.2.3.1 Materials and Functions of Filters

The filter media ranges from mesh screens to depth style media such as threads or chopped paper to 100% natural cellulose to 100% man-made fibres to almost any conceivable combination in between. Table-1 outlines an overview of filtration applications in the automotive industry.

Table-1: Automotive filtration applications and filter media	
Application	**Type of filter media**
Carburetor air filters	Mainly nonwoven (wet, dry, needled or spun bonded)
Engine oil filters	Resin impregnated wet laid nonwoven (paper)
Fuel tank filters	Activated carbon
Cabin interior filters	Electrostatically charged fibre media, nonwovens, activated carbon, specialty paper
Diesel/soot filters	Ceramic materials
ABS wheel/ brake filters	Metal or fibre woven screens
Power steering filters	Mainly screen fabrics
Transmission filters	Woven fabrics or needle felts
Wiper washer screen filters	Woven fabrics
Air conditioning recirculation filters	Nonwoven / activated carbon
Crank case breathe filters	Nonwovens

With so many media choices, it becomes a complex aspect to choose a right kind of media for an engine system. Typically, if only larger particles are to be removed, a very basic cellulose media is used. As the size of contamination to be removed gets smaller and smaller, the type of media changes from more complex cellulose to blended media where cellulose and man-made fibres are blended together in various configurations. For the removal of extremely small contamination, media typically changes from one dominated by cellulose to one made exclusively from various types of man-made fibres. The logic behind choosing man-made fibres is owing to its level of achievable fineness. Finer the fibre greater will be the particle capturing. Recently nano-fibre materials are becoming popular.

1.2.3.2 Air filters

Air is vital to vehicles engine. It is mixed with fuel, ignited and with the resulting controlled explosion provides power to vehicle. It takes between 10,000 to 12,000 gallons of air for each gallon of fuel. The only way for air to enter the engine is through the air intake after passing through the air filter (Fig.1), which is essential in removing contaminants such as dirt particles, dust and debris from the air penetrating the engine, which can cause damage to engine cylinders, wall, pistons and piston rings whilst allowing high volume of air to pass through. If the air filter is clogged up, engine performance is reduced, engine power decreases and engine wear is increased.

An established site for the air filter in the engine of an automobile is in the carburetor. The filter stationed here is normally a wet laid nonwoven material. Commonly this medium is a mixture of cellulose fibres that are derived from the wood pulp, with small amount of synthetic fibre. The fibre scaffold is imparted with stiffness by a resin binder. Air filter can consists of 1-20 layers, which are normally pleated to increase the surface area. In multilayer, the densest layer resides on the exit plane and a less dense layer covers the entry. This arrangement confers a gradient density capability where by gradually smaller particles are held in the deeper layers of the filter. Another version of this medium is available in which the fibres are more intermingled.

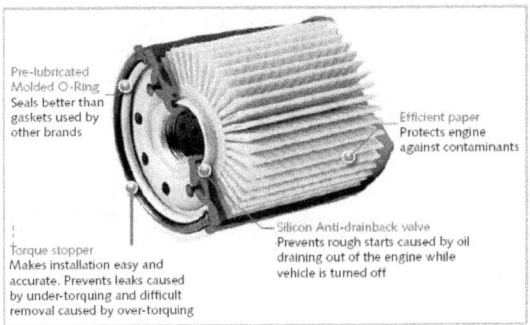

Apart from these two techniques, it is also possible to manufacture filter by dry laying. Generally, this kind of filter consists of a fleecy top web, an open structured inner layer, and a dense lower layer. This type of filter is composed of a blend of synthetic and cellulosic fibres. Inside the filter the cellulosic fibres normally furnish the lower section, which is particularly dense. At present, most development on the carburetor air filter is being conducted in Japan, where Toyota have designed a more expensive media that is made from needled felt and spun bonded material

Filtration through an engine mounted air filter is accomplished through four mechanisms. The first mechanism is where foreign particles adhere to the media via the impact impingement mechanism. Here, dirt strikes the filter and attaches itself permanently. In the second mechanism, dirt can collect as cluster aggregates, which are collection of particles, created by the triboelectric nature of fibres or the electrostatic charge that is generated during fluid flow. Thirdly, dirt can be trapped between the pores of the filter media, particularly when they become clogged and reduced in size. The last opportunity for a filter to capture dust arises from the interaction between fluid flow and pressure drop. When all these mechanisms are employed, the total void volume of an air filter can be filled to 90% by the end of its useful life.

1.2.3.3 Oil Filters

Dirt is one of the major causes of engine wear. Dirt particles are extremely abrasive and these particles are carried by the oil into the precision clearances between bearings and other moving parts. Once they work in between in these parts, they grind and groove surfaces, altering clearances, and generating more debris that is abrasive. As this wear cycle continues precision components become progressively sloppy and fatigued, until they fail altogether. In addition to physically assaulting engine components, dirt and other contaminants work to degrade the oil that provides vital engine lubrication. Dirty carbon particles generated during combustion can be forced past piston rings and into the oil. These particles by their very nature act like tiny sponges, absorbing critical additives, thus shortening oil life.

Further in the presence of moisture, common by products of combustion will react chemically to produce corrosive and rust producing acids. Typically about 80% of engine wear is due to contaminants <10 microns. The function of oil filter is to remove soot, rust particles and other solid contaminants from the oil, providing maximum protection and safety to engine (Fig. 2). The Figure 3 shows the oil filter technology. The earliest incarnations of these filters were made from woven, mesh and paper media that caught dirt by surface phenomenon. Filters made from polyester needle felts were impregnated and is capable of trapping contamination inside its structure. In an important development nanofibre oil filters are made with premium advanced synthetic media technology that results in fibres that have a controlled size, shape and smaller fibre diameter. The controlled media manufacturing process, low density, greater surface area, and tight pore size allows nanofibre oil filters to deliver both higher dirt holding capacity at the same pressure differential and higher efficiency compared to conventional cellulose filters.

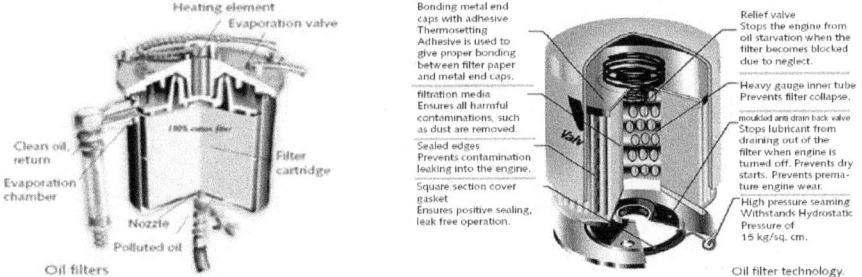

Cellulose fibres are inconsistent in size and shape, allowing more contaminants to pass through, resulting in higher restriction and lower capacity. The synthetic media also has better durability with usage. Throughout the service life of a cellulose filter, hot oil will degrade the resins that bind the media. The synthetic media technology uses a wire screen backing pleated with the media, resulting in superior strength. Nanofibre oil filters offer extended service intervals, greater engine protection to prolong engine and equipment life, improved lubricant flow, improved cold start performance and lower operating costs.

1.2.3.4 Fuel Filters:

Fuel is volatile petroleum liquid that is used to power the engine. The fuel that enters the engine must first pass through the fuel filter as shown in the below figure, which is essential in helping to protect fuel system components such as plug fuel injectors, carburetors etc from contaminants that may be present in the fuel. In extremely cold climate, they are needed to prevent engine trouble caused by water freezing in the fuel.

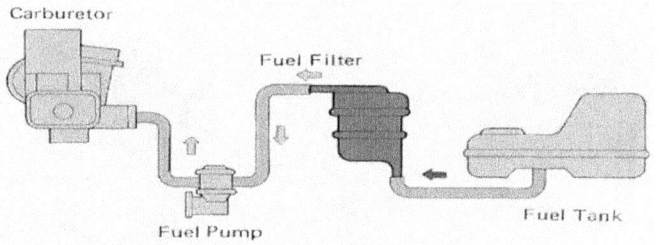

Passage of fuel

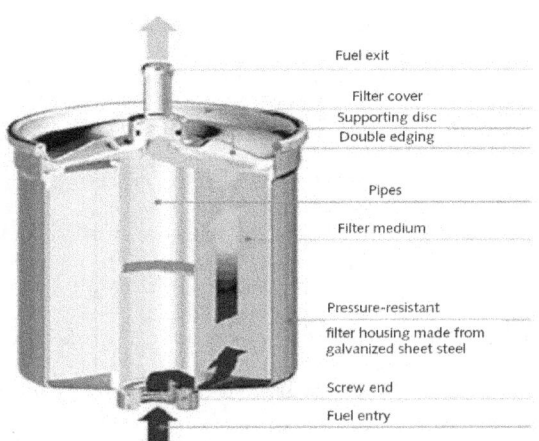

	Fuel exit
	Filter cover
	Supporting disc
	Double edging
	Pipes
	Filter medium
	Pressure-resistant
	filter housing made from galvanized sheet steel
	Screw end
	Fuel entry

Fuel Filtration Technology

Contaminants	Problems
Particulate & Debris	Enters when fuel is transferred between storage tanks. Particulate in the fuel can disrupt engine combustion and cause wear on injectors.
Water	Water in the fuel causes corrosion and reduces the lubricity of fuel. It can negatively affect the combustion process and consequently damage system components. Water enters the fuel from storage tanks.
Wax/Paraffin	Drop out of fuel in cold weather conditions.
Microbes (Bacteria)	Can grow in the water at the fuel interface.
Fuel Degradation Products (FDP)	Fuel by-products result from the thermal and oxidative instability of fuel prior to combustion
Asphaltenes	Found naturally in crude oil and can be found in refined fuel.
Air	Enters the system from leaks in the fuel line or system connections.

When an engine receives less contamination it exhibits higher power, greater fuel economy, and lower emissions. Contaminations can originate from dirty and rusty service station storage tanks and, as the vehicle ages, from corrosion within the fuel system components. Water, fungus, bacteria, wax, asphaltite, sediment and other solids are the major contaminations present in the fuel. Water is the greatest concern because it is the most common form of contaminant. This is likely to be introduced into the fuel supply during fueling when warm; moisture laden air condenses on the cold metal walls of fuel storage tanks or from poor housekeeping practices. The effects of water in fuel can be serious. Water can cause a tip to blow off an injector, or reduce the lubricity of the fuel, which can cause seizure of close tolerance assemblies such as plungers. Further the microorganisms (also called humbugs) live in water and feed on the hydrocarbons found in fuel, which cause plugging of a fuel filter through multiplying colonies spreading throughout a fuel system.

The fuel filter will have a slime coating over the surface of the media, dramatically reducing the service life of the filter. Bacteria may be any color, but is usually black, green or brown. Draining the system will reduce microbial activity, but will not eliminate it. The only way to eliminate microbial growth once it has started is to clean and treat the system with a biocide. Wax while desirable as a source of energy in fuel, but its control in cold weather operation is needed. Wax crystals form as a result of cold temperature precipitation of paraffin. Temperatures below a fuel's cloud point will result in wax precipitation and filter plugging. To prevent plugged filters due to wax formation, the cloud point of fuel must be at least +12°C (+22°F) below the lowest outside temperature.

Fuel suppliers blend diesel fuel based on local anticipated cold weather conditions. Particular attention should be given to fuel purchased outside one's own area since it may not be suitable for operating conditions depending on climatic variation. Asphaltites are components of asphalt that are generally insoluble and are generally present to some extent in all fuel. These black, tarry asphaltites are hard and brittle, and are made up of long molecules. Fuel with a high percentage of asphaltites will drastically shorten the life of a fuel filter. Sediment and other solids often get into fuel tanks and cause problems. Most of the sediment can be removed by settling or filtration. Fuel filters designed for specific applications will remove these harmful contaminants before they cause further system wear and damage.

Engine fuel filter media is most commonly a pleated cellulose base material. This fuel filter media is tested for compatibility with a variety of diesel fuels, including biodiesel. Larger particulate are trapped on outer layer, while finer particles are held deeper in the media. Treating a cellulose media with a silicon based treatment allows for effective water separation. Typically, this media is used on the suction side of the fuel system to remove harmful water and coarse particulate contaminant. Water coalesces on media and drains to bottom of can/bowl. Particulate is trapped and held in media. The third generation fuel filter water separator media uses both cellulose and melt blown synthetic layers to achieve the highest levels of fuel filtration performance. This double-layered media increases particulate holding capacity and is a high performance water separator. It has the ability for high efficiency emulsified water separation and can be used in both suction and pressure sides of fuel systems. The polyester layer improves water separation and dirt holding capacity performance.

The first petrol filters were made from wire mesh and, although they were efficient, they were unable to separate water from passing fuel. New filters based on vinylidene chloride polymers (namely Saran® monofilament) are produced by various companies under different trade name. The filaments are made by melt spinning and are then stretched. Some of their significant properties are resistance to water, fire, light, and bacterial attack. Saran® is a prime material for petrol filters and are resistant to automotive fuel, deliver high mechanical strength and recovery, flame retardant, and do not absorb water. In addition, the Saran® filter is able to prevent the ingress of air in to the fuel tank. The success of the filter is also due to its wicking capability. To facilitate wicking, the surface tension of the Saran® filter fibre is greater than the critical surface tension of the fuel. Wicking ensures that the filter is constantly seeped in fuel in the presence of air and fuel vapour. Indeed, even when the fuel tank is nearly empty, a fine film of petrol collects over the filters surface and prevents air from penetrating its structure. Overall, the wicking function is governed by the filter medium and its construction, chiefly by the choice of fibre, finish, and pore distribution. Most saran filters are fitted in to vehicles that employ a carburetor system. Here, when the engine is running, the flow of fuel is variable and intermittent. The role of the saran filter is to intercept any contamination and prevent it entering its structure or passing into the fuel tank. Particles cannot adhere to its surface as filter cloth is exceptionally smooth. Instead, the filter performs a self-cleaning operation called "back washing", where by impurities are shed from its surface when car stops.

Filter units are sized by flow rate, expressed in gallons per minute or hour. Since a four-cycle diesel returns two to four times as much fuel as it burns and two-cycles return five or six times, a handy rule of thumb is to multiply peak engine fuel consumption by 3.5 or 4, then divide by 60 to get the fuel flow rate in gallons per minute for a four-cycle diesel.

1.2.3.5 Cabin filters

The quality of the air inside the vehicle is a major factor in protecting the health and safety of the driver and passengers. The level of air borne pollution ranges from 0.05-0.5mg /m^3.The particles that are larger than 3 microns can be trapped in human breathing channels and responsible for instigating allergic reactions like hay fever and asthma. The cabin air filter helps trap pollen, bacteria, dust, and exhaust gases that may find their way into a vehicle's ventilation system, making the interior of the car a healthier place.

Cabin filters are generally located under the vehicles hood, inside the glove compartment or under the dashboard and most are within easy reach for quick replacement. When the vehicle is in motion, or when the ventilation system is in use, all impurities that are present in the outside air are sucked into the cabin like a vacuum. The cabin air filter is typically a pleated-paper filter that is placed in the outside-air intake for the vehicle's passenger compartment. The cabin air filters also protect the heating and air conditioning system from pollution, guarantee a clear view and thus contribute to safer driving.

Cabin filters used in the vehicle can be classified as pollen filter, combination filter and particulate filter. The type of cabin filter used is decided by environment conditions and vehicle type. A pollen filter removes spores, pollen, particulates, bacteria, fungi and road dust. Larger particles (> 3μm) are mechanically separated from the airflow by the dense weave of the filter fabric, although extremely tiny particles such as diesel soot may not be removed.

(a) Pollen Filter   **(b) Activated Carbon filter**

A combination filter with a laminated construction consisting of particle filter fabric, activated carbon layer and substrate medium, can absorb both particles like the pollen filter as well as odors and harmful gases. Activated carbon has an extremely porous structure which can be subdivided into different pore sizes. A filter of this type can noticeably reduce and even eliminate hydrocarbon molecules, benzene, odors and other gases. Ozone decomposes almost completely into harmless oxygen when it contacts activated carbon.

Particulate filter is a multilayer design composed of a pre-filter, an electrostatically charged microfibre layer and a cover layer. The pre-filter is made from coarse polyester fibres that are supported by binder. To impart the pre-filter with certain functional properties, the binder is formulated with a mixture of antibacterial, water repellent, and flame retardant agents. Pre-filters are designed to capture larger contaminating particles, including pollen and mold spores. The electrostatically charged microfibre layers, made up of melt blown polypropylene fibres attracts and holds smaller-size particles from smoke, bacteria, and other contaminants. The cover layer

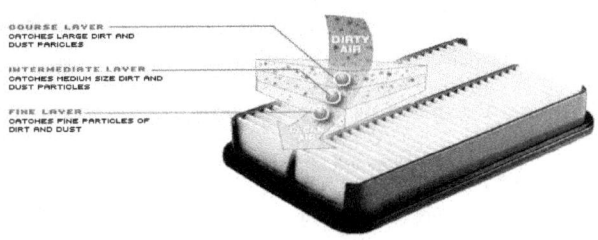

Cabin filter - Particulate filter

is constructed of a nonwoven filter media that adds stability and protects the microfibre layer from damage. It also provides an additional barrier against harmful contaminants. The activated charcoal filter has all of the features of the particulate filter, plus an activated charcoal layer that absorbs harmful gases and their odours.

1.2.4 TYRE CORDS

Tyres are highly engineered structural composites whose performance can be designed to meet the vehicle manufacturers' ride, handling, and traction criteria, plus the quality and performance expectations of the customer. The tyres of a mid-sized vehicle roll about 800 revolutions for every mile. Hence, in 50,000 miles, every tyre component experiences more than 40 million loading-unloading cycles, an impressive endurance requirement. The first practical pneumatic tyre was made by John Boyd Dunlop in 1887. Dunlop is credited with "realizing rubber could withstand the wear and tear of being a tyre while retaining its resilience". Today over 1 billion pneumatic tyres are manufactured in around 450 tyre factories in the world.

The tyre is a complex technical component and must perform a variety of functions. The reinforcing material gives the product the shape and resistance to dynamic stresses and it also determines the service life, loading capacity, abrasion resistance and many other properties of the product. Most important of all, however, it must be capable of transmitting strong longitudinal and lateral forces (during braking, accelerating and cornering) in order to assure optimal and reliable road holding quality. Textiles used in tyre reinforcement are specially prepared and processed fabrics called tyre cord fabrics. Tyres being made of rubber are not dimensionally stable by itself on application of load. Also with the advent of pneumatic tyre the dimensional stability requirement increased as tyres became a pressure vessel. The tyre also acts as a contact between the vehicle and the road and steers the vehicle; the reinforcing textile cord thus monitors directional stability of the tyre.

Vehicles with powerful engines require, for example, good grip–particularly on wet and flooded roads. On the other hand, a corresponding improvement in the tread compound can affect tyre life, rolling resistance and ride comfort. One point, however, has absolute priority over all other tyre design objectives, and that's safety. In addition, tyre play an important role in the overall energy consumption of an automobile, e.g. in a running car tyres uses about 5-6 % of fuel on average.

1.2.4.1 Functions of Tyres:

1. Vehicle to Road interface
2. Support vehicle load
3. Road surface friction
4. Absorb Road irregularities

1.2.4.2 Tyre – Parts:

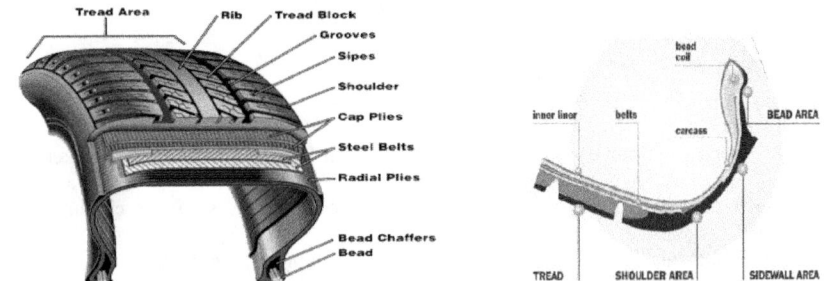

Depending upon the alignment of tyre cords, tyres are classified as bias, radial and belated-bias.

(a) Diagonal (bias) tyres

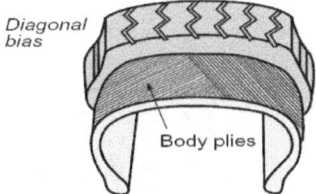

In Bias tyres, body ply cords are laid at angles substantially less than 90° to the tread centerline, extending from bead to bead. These types of tyres are used in trucks, trailers and farm implements. Simple construction and ease of manufacture are advantageous. As the tyre deflects, shear occurs between body plies which generate heat. Tread motion also results in poor wear characteristics.

(b) Belted bias tyres

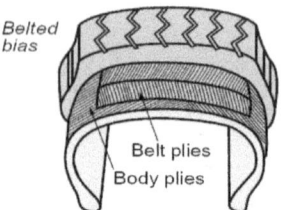

Belted bias tyres, as the name implies, are bias tyres with belts (also known as breaker plies) added in the tread region. Belts restrict expansion of the body carcass in the circumferential direction, strengthening and stabilizing the tread region. Improved wear and handling due to added stiffness in the tread area are advantageous. Body ply shear during deflection generates heat; higher material and manufacturing cost.

(c) Radial tyres

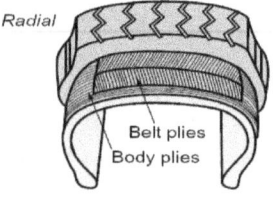

Radial tyres have body ply cords that are laid radially from bead to bead, nominally at 90° to the centerline of the tread. Two or more belts are laid diagonally in the tread region to add strength and stability. Radial body cords deflect more easily under load, thus they generate less heat, give lower rolling resistance and better high-speed performance. Increased tread stiffness from the belt significantly improves wear and handling. Complex construction increases material and manufacturing costs.

1.2.4.3 Reinforcement Materials

The tyre cord and bead wire are the backbone of the tyre. The tyre cord and bead wire are the predominant load carrying members of the cord-rubber composite which determines the load carrying capacity, the comfort of the tyre and the dynamic behavior. Some tyres, when parked, can develop a temporary "set" in the rubber compounds and reinforcement cords, referred to as a "flat spot". Five materials currently make up the major tyre textile usage – rayon, nylon, polyester, aramid, and steel.

Fibres used in Tyre Reinforcement	Advantages	Disadvantages
Nylon 6 and Nylon 6,6	Good heat resistance and strength; less sensitive to moisture	Heat set occurs during cooling (flat spotting); long term service growth
Polyester	High strength with low shrinkage and low service growth; low heat set; low cost.	Not as heat resistant as nylon or rayon
Rayon	Stable dimensions; heat resistant; good handling characteristics	Expensive; more sensitive to moisture; environmental manufacturing issues.
Aramid	Very high strength and stiffness; heat resistant.	Cost; processing constraints (difficult to cut).
Steel cord	High belt strength and belt stiffness improves wear and handling.	Requires special processing; more sensitive to moisture.

(a) Nylon

Nylon 6 and Nylon 6, 6 tyre cords are synthetic long chain polymers produced by continuous polymerization/spinning or melt spinning. Its low modulus and low glass transition temperature make it unacceptable as a belt material or for applications where aesthetics, ride, and handling are important, i.e., in passenger tyres. Nylon is preferred in uses requiring carcass toughness, bruise and impact resistance, high strength, and low heat generation, e.g., in tyres for medium and heavy-duty trucks, off-road equipment, and aircraft. In these applications nylon can be used in the bias-ply tyre carcass or in radial tyre carcasses with steel or aramid belts.

	Rayon	Nylon 6	Nylon 66	Polyester	Aramid	Steel
Tenacity (cN/Tex)	50	80	85	80	190	35
% Elong at break	6	19	16	13	4	2.5
Modulus (cN/Tex)	800	300	500	850	4000	1500
Shrinkage (% at 150C)	<0.1	6.0	5.0	2.0	<0.1	<0.1
Moisture regain (% at RT)	13	4.5	4.5	0.5	<2.0	<0.1
Specific gravity	1.52	1.14	1.14	1.38	1.44	7.85
Melting temperature (C)	>210	225	250	250	>500	--
Glass transition temperature(C)	--	55	55	80	--	--
Heat resistance (C)	150	180	180	180	250	--
Approximate relative cost per unit weight (PET=1.00)	1.33	--	1.13	100	5.00	--

(b) Polyester

Polyester tyre cords are also synthetic, long chain polymers produced by continuous polymerization/spinning or melt spinning. The most common usage is in radial body plies with some limited applications as belt plies. Polyester is superior to nylon tyre cords in some respects e.g. less thermal shrinkage, less flat spotting tendency, but it suffers lack of bonding with rubber. Polyester cord is not recommended for use in high-load/high-speed/ high-temperature applications, as in truck, aircraft and racing tyres, because of rapid loss in properties at tyre temperatures above about 120°C.

(c) Rayon:

Rayon tyre cords are made from cellulose produced by wet spinning. It is often used in Europe and in some run-flat tyres as body ply material. The low- shrink, high-modulus, good-adhesion properties of rayon make it an excellent choice for use in passenger tyres. However, rayon has lost market share to polyester due to higher cost and environmental concerns with production facilities. Rayon is used for racing tyres and has gained renewed interest in the development of an extended-mobility self-supporting passenger tyre.

Dimensional stability (Carcass)	
Uniformity in curing	Rayon > Advanced Polyester >Nylon
Appearance (sidewall indentations)	Rayon > Advanced Polyester >Nylon
Dynamic stiffness (steering)	Rayon > Advanced Polyester >Nylon
Flat spotting	Rayon > Advanced Polyester >Nylon
Durability (Carcass)	
Fatigue resistance/heat generation	Nylon > Advanced Polyester > Rayon
Impact Resistance (Toughness)	Nylon > Advanced Polyester > Rayon
High speed/ run-flat tires	Rayon > Advanced Polyester
Strength	Aramid > Steel > Nylon > Polyester > Rayon
Modulus(stiffness)	Aramid > Steel > Rayon >> Polyester > Nylon
Elongation	Nylon > Polyester > Rayon > Steel = Aramid
Compression fatigue	Nylon > Polyester > Rayon > Steel > Aramid
Chemical resistance	Aramid > Nylon > Rayon > Polyester > Steel

(d) Aramid

Aramid is a synthetic, high tenacity organic fibre produced by solvent spinning. Aramid cords have very high strength, high modulus, and low elongation. It is 2 to 3 times stronger than polyester and nylon. It can be used for belt or stabilizer ply material as a light weight alternative to steel cord. It is particularly suited where weight is important, such as in the belts of radial aircraft tyres or in overlay plies for premium high-speed tyres. In a multiply carcass construction, aramid's low elongation will prevent the outer ply from adjusting to the average curvature, thus placing the inner plies into compression.

(e) Steel Cord:

Steel cord is carbon steel wire coated with brass that has been drawn, plated, twisted and wound into multiple-filament bundles. It is the principal belt ply material used in radial passenger tyres. Bead wire is carbon steel wire coated with bronze that has been produced by drawing and plating. Filaments are wound into two hoops, one on each side of the tyre, in various configurations that serve to anchor the inflated tyre to the rim.

1.2.4.4 Tyre Cords:

Tyre cords are built up from yarns which in turn come from filaments. Filaments from production spinnerets are gathered together, slightly twisted, and placed on "beams" for further processing. The filaments are twisted "Z" into yarns and the yarns are back twisted "S" to form a cord. The size of a tyre filament, yarn, or cord is measured by its linear density or "denier" or "decitex". Thus a 940/2 8x8 nylon cord is formed from 2 - 940 decitex yarns twisted separately at 8 turns per inch and then back-twisted together at 8 turns per inch to form the cord. A 1650/3 10x10 rayon cord would comprise 3-1650 denier yarns twisted at 10 turns per inch separately and back-twisted together at 10 tpi.

As with all tyre components, choice of a textile cord for a given tyre application may require compromises involving cost, intended market segment and end-use application. The tyre engineer has a number of choices for a tyre textile:

- Chemical composition of textile
- Cost per unit length and weight (cost in tyre)
- Denier – filament size and strength
- Cord construction – number of yarn plies
- Cord twist

- Number of cords per unit length in ply

- Number of plies in the tyre

Tyre performance	Related property of Tyre cord
Bursting strength	Tensile Strength
Tyre Endurance	Adhesion with Rubber
Power Loss	Viscoelastic Properties
Tread Wear	Modulus
Tyre Size & Shape	Modulus
Tyre Groove Cracking	Creep
Flat Spotting	Thermal Shrinkage
High Speed Endurance	Heat Resistance

Twist levels are important for tire cord performance. Higher twists allow a cord to behave like a spring which will not open up under compression, while lower twists allow a cord to behave as a rod, maximizing the strength. As twist increases the tenacity decreases, fatigue in compression improves (the main reason for higher twists), the cord cost per tire increases (because cords become shorter as they are twisted), and shrinkage during processing and cure increases. Tenacity and fatigue resistance are sometimes reduced with increasing twist. The three major system used to form the cord are ring twisting, direct cabling and two for one twisting (TFO).

Cord Property changes with increasing twist levels
Reduction in strength
Reduction in initial modulus
Reduction in cyclic tension fatigue resistance
Increase in elongation to break

Increase in rupture energy

Increase in cyclic compression fatigue resistance

Increase in cord cost per tire

Functions of Tyre cords
• Maintain durability against bruise and impact
• Support inertial load and contain inflating gas
• Provide tyre rigidity for acceleration, cornering, braking
• Provide dimensional stability for uniformity, ride, and handling

Cord Requirements
• Large length to diameter ratio, e.g., long filaments
• High axial orientation for axial stiffness and strength
• Good lateral flexibility (low bending stiffness)
• Twist to allow filaments to exert axial strength in concert with other filaments in the bundle
• Twist and tyre design to prevent cord from operating in compression.

Ideal Cord properties for a Tyre Carcass
Dimensional stability –low shrink during cure, no flat spotting, no long term growth
High tensile strength
High tensile modulus
Low bending modulus

High durability – fatigue resistance, low heat generation on flexing, high adhesion to rubber, Chemical and oxidation resistance, heat resistance

High toughness – impact and abuse resistance

Low hysteresis loss at high speeds

1.2.4.5 Weaving of Tyre fabrics:

Tyre fabrics have changed in response to the constant demand for better tyre performance. The cords are woven, with pick cords to maintain spacing, into a wide sheet of fabric. The function of the pick is to maintain a uniform warp cord spacing during the downstream operations, such as, shipping, adhesive dipping and heat treating, calendering, tire building and lifting. Usually rapier and shuttle looms are used for weaving cord fabric, nevertheless air jet machines from different companies have been specially designed for tyre cord weaving. Uniform cord distribution in the finished tire is essential for tire uniformity and performance. The fabric is then passed through a four-roll calender where a thin sheet of rubber (body ply skim) is pressed onto both sides and squeezed between the cords of the fabric. The calendered fabric is wound into rolls, with a polypropylene liner inserted to keep the fabric from sticking to itself, and then sent to a stock-cutting process.

1.2.4.6 Heat Treatment

The heat treatment is mainly applied to the woven fabrics made of thermoplastic synthetic fibres, such as nylon and polyester, to be incorporated in tyres. The heat treatment occurs under controlled time (exposure), tension and temperature conditions. It helps to optimize the fabric properties, especially improving the dimensional stability of the fabric and the overall setting effect.

1.2.5 Automobile Interiors

1.2.5.1 Automotive Seats

a) Seat Comfort:

During a journey, vehicle occupants are subjected to both mental and physical stresses when exposed to road vibrations, dense traffic, noise and different weather conditions. Human exposure to mechanical vibrations causes fatigue and discomfort. The magnitude of the effect depends on the intensity, duration and directions of the excitation. Seat is one of the main aspects to be considered when defining comfort in a moving vehicle. Comfort on automotive seats is dictated by a combination of static and dynamic factors. Term "comfort" is used to define the short-term effect of a seat on a human body; that is, the sensation that commonly occurs from sitting on a seat for a short period of time. In contrast, the term "fatigue" defines the physical effect caused by exposure to the seat dynamics for a long period of time. Comfort is subjective and it is difficult to define this term objectively in order to determine design specification of seat that will provide this attribute to an occupant (Pywell, 1993). Seating comfort is strongly related to physical comfort of an occupant. Physical comfort can be defined as the physiological and psychological state perceived during the autonomic process of relieving physical discomfort and achieving corporeal homeostasis. There are three modes of comfort identified; static, transient and dynamic comfort (Shen and Vértiz, 1997). Homeostasis is a state of equilibrium between different but interrelated functions or elements, as in organism or group.

b) Seat Materials:

The construction of the automotive seat and the cushioning material employed to provide the occupant a comfortable place to sit while operating a motor vehicle have continued to evolve. Most early autos had large, padded seats essentially similar to furniture seating. Some of the materials employed in the 1920s in auto upholstery are: springs, hair and its substitutes, cloth, duck, sheeting, cambric, muslin, buckram, webbing, cotton wadding and battings, down, feathers, top materials, slip cover materials, coated ducks, cords and bindings, etc. In order to ensure the hair fibres did not disintegrate or loosen from the seats, latex-bonded seating interiors were developed. A light impregnation with natural or synthetic rubber latex bonds the fibres together where they touch. Rubberized hair pads have been described by Pole in 1959.

Rubberized hair automotive pads

However, Mercedes® used fibre/latex seating for many years since it apparently gives their seats the desired performance of breathability and good lateral support cornering especially at high speeds. Rubberized coconut fibre seating was used by VW in Brazil up until 1994. Latex foam seating (without any reinforcing fibres) has been used for decades. Latex foam was considered a much superior, comfortable product to previous types of seating, e.g., horsehair, wadding, springs. Latex foams were considered the comfort yardstick. They were replaced by polyurethane foam types that for many years did not equal latex foam in comfort. They can be considered for two major reasons.

1. The urethane process is a easy process to master and run in production to obtain consistent product, easily trimmed into car seats.

2. The cost of producing urethane cushioning was lower than the more complicated latex processes.

One of the major advantages of PU foams over latex foam is their resistance to bottoming-out when stressed. Latex foam has inferior UV resistance compared with PU foam and latex may show the early formation of a heavy crust whereas PU foam may exhibit some surface crumbling and/or friability after long UV exposure. DuPont® was able to make successful foam using their "Teracol®" (polybutylene ether) products. It had the extra resiliency required without the bad creep of the current polyester-based flexible PU foams. Its major drawback was that it was too expensive so other new materials had to be found. These can be generally classified as polyether polyols such as:

  i) Polypropylene glycols (UCC, Dow)

  ii) Polypropylene-polyethylene glycols (UCC – UCONS)

  iii) Pluronics – Wyandotte (now BASF)

High Resilience (HR) Polyurethane molded foams are key components of automobile interiors and contribute to passenger comfort especially in seats. They have superior vibration damping capability over a wide range of frequencies at low material density. HR foam had a latex rubber-like feel, higher support ratio, lower hysteresis loss, good flex fatigue improved inherent flammability resistance. Some special PET nonwovens have recently been developed such as a fibre mass stabilized with elastomer fibre-fused bonding, a folded web and stitch-bonded web, and a PET three dimensional knit fabric with super water absorbency. One of the most important advantages of this fibrous material over urethane foam is their good moisture permeability, which keeps passenger's physiological comfort.

Important requirements of Car Interior
Good management of heat and water vapour transfer;
Anti-stain/easy to clean characteristics;
Antimicrobial/antibacterial properties;
Anti-allergic trimming;
Flame resistance;
Antistatic properties; and
Improved acoustic performance

The traditional seat covering used in the vehicles in the early days was leather and the use of a textile fabric was rare and constituted a surcharge. In today's modern vehicles the role of textile fabrics are more than leather coverings. Leather was used for vehicle interiors because of its durability and strength regarding wear, despite the cost involved. The use of textile fabric was considered but the fabrics produced would have been basic and inadequate for the task of an interior covering for a vehicle. Today's involvement of the textile industry in the automotive industry is considerable. The essential property requirement of seat cover laminates are aesthetic effect, physiological conformability, strength, color fastness, flame retardancy, heat resistance, and nonvolatile substance content. They are also often required to have some special functions such as antibacterial, antistatic and stain-resistant properties.

Polyester fibre is predominantly used for making seat covers and foursome seat cushion material. It is also used for making door trim material because polyester fibre has a higher modulus, good heat stability, excellent resistance to color change, and high durability for sunlight degradation and is less expensive. Nonwoven polyesters fabrics made from recycled fibres and novel knitted structure such as spacer fabrics, kunit, multiknit and struto have been considered as substitutes for polyurethane foam in the cover laminate. Spacer fabric is a knitted textile with threads running perpendicular to the plane of the fabric with a knitted layers each side.

Spacer Fabric Structure

Multiknit comprises two stitch layers with the pile in between, which makes fabrics from fibrous webs using Malimo knitting techniques. Kunit consists of a stitch layer with a pile on the top.

The seat covering fabrics are produced with fabric structures such as pile weave, tricot with raising, pile double raschel knit, and pile circular knit. Pile fabrics are usually related to an increase in the values of tactile and visual aesthetic effects. General trends are towards an increase in knit fabrics (tricot, double raschel, and circular knit) that are less expensive and have more formability than weave fabrics. Recently, a suede fabric using PET microfibre nonwoven as base material has been introduced. Seat cover containing phase change material (PCM) have also been developed, which can increase the microclimate comfortability of the seat.

1.2.5.2 Door Trims, Roof Trims and Floor Coverings

Door trims are usually made up of plastics such as vinyl chloride and polyolefin. But textile materials are also used for higher-class- cars. In most cases, the textile material is the same as the seat cover fabric. However, the fabric needs to have high enough formability to be made into the complicated shape of a door trim. Its lower end is usually covered by the same carpet as the floor because the door often gets kicked. The fabrics used for door trims should have high degree of stretchiness. This can be obtained by incorporating elastomeric yarns into structure of fabrics. Now a layer of foam is added to the door panel fabrics to get a more luxurious touch.

Coconut fibres have been used in cars for more than 60 years in such applications as interior trim and seat cushioning. Unlike plastic foam, the coconut fibres have a good 'breathing' property which is a distinct advantage for vehicle seats being used in countries where the climate is hot, as is the case in Brazil. Coconut fibres are also naturally resistant to fungi and mites and the remains of the fibres also make an effective natural fertilizer at the end of their lifetime. In 1994, Daimler Chrysler started using fl ax and sisal fibres in the interior trim components of its vehicles. They continued investing in their application of natural fibres and in 1996 Jute was being utilized in the door panels of the Mercedes Benz E- Class vehicles. German Car manufacturers are aiming to make every component of their vehicles either recyclable or biodegradable

PET nonwoven and tricot are used as a roof trim. The use of needle punched nonwoven in particular has increased and especially velour patterned nonwoven is usually formed into roof trim by integration with base material. Trims need to have good resistance to colour fastness for sunlight, heat resistance, mechanical durability, light durability, formability, non-volatile substance content and strain resistance in addition to lightness. Sound absorbing capability and heat insulation are also needed. The base material for roof trim can be mainly classified as polyester foam sheets and fibre reinforced porous polymer sheets .Glass fibre is usually used for reinforcing these sheets.

Floor coverings are made up of tufted cut-pile or tufted loop-pile or needle-felt. The uses of nonwoven carpet have significantly increased because it is more economical and it has better formability. Road noise is considered as an environmental pollution in few countries. Carpets by providing thermal and acoustic protection thus directly contribute to safety. The absorption potential of a tufted carpet as well as the pile density and the yarn are to be taken into account besides the thickness of the absorbing layer.

1.2.6 Other Textile applications in Automobiles

a) Head liners

The headliner plays a number of roles within the car's design and engineering; its first role is to cover the inside of the metal panel of the roof. As well as covering the metal, the unit also hides wiring and curtain airbags. They also act as carriers for other components such as storage boxes and overhead lighting. Important requirements of head-liners are lightweight, thin profile but rigid without any tendency to buckle, flex or vibrates directional stability, aesthetically pleasing, soft in touch.

Headliner consists of two core materials, the substrate (which can be something as basic as cardboard, but it is normally polyurethane or another polymer impregnated with glass fibres for strength and stiffness) and the facing fabric. In European vehicles, polyurethane-based headliners are used while in North America, PU is joined by polyester (PET) and a range of new materials. As well as covering the roof panel and acting as a carrier for components with the overhead system, the headliner assembly also has an important acoustic function; either built-in to its make-up or design or through acting as a cover for acoustic pads or filler materials which are fixed to the inside of the car's roof.

b) Trunk Liners

Trunk liners gives a most luxurious appearance and also serves as protection to both exterior walls and to trunk content such as luggages perfect to date. The liners must be decorative and functional, yet have relative cost. These are usually made from waste fibre that are needled and then naturated with elastomeric materials. However, even spun bonded polypropylene is also used as a substrate for a foamed rubber material that is used as trunk liner.

c) Parcel Shelves

Parcel shelves, also referred to as package trays are now almost invariably covered with needle-punched nonwoven mainly in polypropylene or polyester.

d) Dash board

The dashboard, probably the hottest area in the car interior, offers some opportunity for textiles. The dashboard shape being highly curved and also complex, only knitted fabric, and 3D knitted fabrics would be eminently suitable. The performance requirements of textiles used for dash board are: low gloss (no glare or reflections on the windscreen), soft touch, pleasant aesthetics, non-fogging, non-odorous, UV stability, resistance to heat ageing, resistance to low temperature, high abrasion resistance.

e) Sun visors

Sun visors are produced from raised -warp knit fabric and PVC. Injection moulding produces some sun visors, others are composed of metal frames and rigid foam or cardboard are also used. The article is close to the windscreen and UV light and heat resistance must be of the highest standard.

1.3 Nonwovens in Automotive applications

Nonwoven fabrics are increasingly used in the automotive industry, while the use of woven and knitted fabrics in automotive applications shows only minimal growth. Nonwoven fabrics offer a better price/performance ratio than textiles made out of yarns for many applications. Nonwoven fabrics are likely to become most widely used in the automotive sector as the trend towards manufacturing lighter weight cars with lower fuel consumption continues. Higher productivity and lower production cost of nonwoven production are the reasons attributed for this. To meet the increasing demand for automotive nonwovens, enhancement of the production capacity of high quality and versatile nonwoven materials is being intensively focused on by nonwoven manufacturers.

1.3.1 Nonwovens used in Cars:

1. Door lining: edge trim, door mirror, arm-rest, lower part (door pocket)

2. Sun visors

3. ABC-pillar covering (covering of seat belt)

4. Headliner (moulded roof): roof insulation, sun roof (cover), hood, hood padding

5. Parcel shelves: speaker covering

6. Boot lining: floor mat, sides (wheel casings), rear cover, back seat wall, spare wheel case

7. Filters: air filter, cabin filter, fuel filter, oil filter, during car manufacture, (lacquering) etc.

8. Engine housing: bumper felts, bonnet lining, rear side, (dashboard), battery separators, and other insulation points

9. Instrument panel: (insulation), instrument panel (lower part)

10. Dashboard mat

11. Seats: lining for backs of seats, laminated padding for seat covers and bottom of seats, upholstered wadding, upholstery cover, reverse sides, head-rest cushioning, seat sub-padding, foam reinforcement, padding for centre arm-rest

12. Floor mats with tunnel: cladding, sub-upholstery (insulating material, stuffing)

13. Interior rear wall lining: floor of the car body, under the back seats (exterior wheel-case)

14. Estate cars and convertibles: side wall covering (lining for the wheel case), boot floor, lining for the hood-case, cover for the hood-case

1.4 Natural/Biodegradable fibres in Automotive Textiles

The recent increase in consumer environmental awareness, along with increased commercial desire to use "greener" materials, has led to new innovations. A study conducted in 1999 indicated that up to 20 kg of natural fibres could be used in each of the 53 million vehicles being produced globally each year. This means that for each new model of car there would be a requirement of between 1,000 and 3,000 tonnes of natural fibres per annum, with some 15,000 tonnes of flax being used in 1999 in the European automotive industry alone. A study by the Nova Institute in 2000 reviewed market possibilities for the use of short hemp and flax fibres in Europe. In this study, a survey of German flax and hemp producers showed that 45% of hemp fibre production went into automotive composites in 1999. One of the attractions of hemp, as compared with flax, is the ability to grow the crop without pesticide application. The potential for fibre yield is also higher with hemp. The type of natural fibre selected for manufacture is influenced by the proximity to the source of fibre, thus panels from India and Asia contain jute, ramie and kenaf, panels produced in Europe tend to use flax or hemp fibres, panels from South America tend to use sisal and ramie.

Virtually all of the major car manufacturers in Germany (i.e., Daimler-Chrysler, Mercedes, Volkswagen Audi Group, BMW, Ford and Opel) now use natural fibre composites in automotive applications. Ford uses from 5 to 13 kg (these weights include wool and cotton). The car manufacturer, BMW, has been using natural materials since the early 1990's in the 3, 5 and 7 series models with up to 24kg of renewable materials being utilized. In 2001, BMW used 4,000 tonnes of natural fibres in the 3 series alone. Here the combination is 80% flax with 20% sisal blend for increased strength and impact resistance. The main application is in interior door linings and panelling. Wood fibres are also used to enclose the rear side of seat backrests and cotton fibres are utilized as a sound proofing material. In 2000, Audi launched the A2 mid-range car which was the first mass-produced vehicle with an all-aluminium body. To supplement the weight reduction afforded by all aluminium body, door trim panels were made of polyurethane reinforced with a mixed flax/sisal mat. This resulted in extremely low mass per unit volume and the panels also exhibited high dimensional stability.

Recently, in the last few years, Volvo have started to use Soya based foam fillings in their seats along with natural fibres. They have also produced a cellulose based cargo floor tray – replacing the traditional fl ax and polyester combination used previously which resulted in improved noise reduction

Component	Weight (Kg)
Front door liners	1.2-1.8
Rear door liners	0.8-1.5
Boot liners	1.5-2.5
Parcel Shelves	< 2
Seat Backs	1.6-2
Sunroof shields	< 0.4
Headrests	2.5

Natural fibre usage per Component

The two most important factors now driving the use of natural fibres by the automotive industry are cost and weight, but ease of vehicle component recycling is also an ever increasing consideration to meet the requirements of the end of life vehicle directive.

1.5 Nanotechnology in Automotive textiles

Nanotechnology promises to benefit many different aspects of industry. Nanotechnology is defined as the precise manipulation of individual atoms and molecules to create layered structures. Nano-size particles can exhibit unexpected properties— different from those of the bulk material. The basic premise is that properties can dramatically change when a substance's size is reduced to the nanometer range. Incorporating nanomaterials into a textile can affect a host of properties, including shrinkage, strength, electrical conductivity and flammability. Nano fibres can be defined as fibres with a diameter of less than 1mm or 1000nm. Nano fibres are characterized as having a high surface area to volume ratio and a small pore size in fabric form. Nano-composite fibres are produced by dispersing nano size fillers into a fibre matrix. Nano fillers can be distributed in a polymer matrix through either a mechanical or chemical process. Depending on the kind of nano material used and the amount and distribution of the nano material - the mechanical, electrical, optical or biological properties of the textile can be altered.

The automobile industry's overarching development goals are to improve fuel consumption, environmental impact, safety, and comfort, as continually growing car use conflicts with environmental pressures and infrastructure limits. Nanotechnology will undoubtedly play a huge role in the way automotive manufacturers deal these changes. The automobile industry is potentially a major beneficiary of nanotechnology developments which promise enhancements and benefits at various levels providing lighter, stronger, harder materials, improved engine efficiency, reduced fuel consumption, improved safety, reduced environmental impact, and comfort. Textile materials have an essential role here, for their use spans from interior door panels, seats materials and paddings, dashboard, to cabin roof and boot carpets, headliner, seat belts, airbags, various filters, tyre cord, and trimmings. Traditional textiles used for automotive interiors face several major challenges such as protection from dust and dirt, ventilation, durability and wear, and fire resistance; all of which call for new high-tech textiles providing enhanced functionalities. Nano enabled textiles may have novel solutions and address multiple functional requirements.

1.5.1 Applications of Nanotechnology in Automotive Textiles

Polymer nano-composites find use in many automotive applications to achieve specific functional requirements. Toyota Motors used N6/clay nano-composites for making the toothed belt cover which exhibited good rigidity and excellent thermal stability. The weight saving was up to 25% due to the lower amount of clays used. N6 nano-composites have also been used as engine covers, oil reservoir tank and fuel hoses in the automotive industry because of their remarkable increase in heat distortion temperature and enhanced barrier properties together with their mechanical properties. To reduce noise within the car, a nanofibrous layer of non-woven polyvinyl alcohol layered on fibrous underlay materials may be used. Compared to conventional materials this offers improved noise reduction, while also providing good heat insulation and weight reduction. Nano clay/polypropylene nano-composites are used for seat back in Honda Acura. Thermoplastic polyolefin nano-composites doors were developed by Chevrolet Impalas. The weight reduction of polymer nano-composites can have a significant impact on environmental protection and material recycling.

Polycarbonate (PC) is much lighter than standard glass, is safer, and is a more flexible material to design with, but physical limitations such as its poor scratch resistance and low UV shielding have prevented its widespread use. Addition of a transparent, scratch-resistant coating containing silica nanoparticles alleviates many of these issues.

1.5.2 Future Scope of Nanotechnology in Automotive textiles

Many researches are going on in the automotive field in bringing nano-enabled textiles for various applications in automobiles. Coating textiles with a hydrophilic coating of TiO_2 nano particles or by plasma treatment may provide good moisture wicking and transpiration absorption that are valuable for comfortable driving. Nano-composites containing organically modified clay dispersed in selected polymer matrices have attracted considerable attention for imparting flame resistance. Novel filters using nano fibres have the potential to offer superior capabilities compared to conventional products such as a large surface area while maintaining porosity to remove harmful substances and greatly improve air quality and safety. These nano filters can be used in air filters and cabin filters in automobiles which will provide high efficiency in particulate filtration. Car interior application will mostly deal with comfort issues – dirt-repellent and antimicrobial textiles and surfaces, nanoparticulate air filters. The application of silver (Ag) nanoparticles to the textile surface will impart antimicrobial or antibacterial properties that could improve hygiene within the car interior and contribute to eliminating unpleasant odours.

Synthetic fibres possess poor anti-static properties, but fabrics containing electrically conductive nano-sized materials, like titanium dioxide (TiO2), zinc oxide (ZnO), antimony doped tin oxide (SnO), have been proved to be effective in dissipating accumulated static charge. Nanoparticles such as SiO_2, Al_2O_3, ZnO and CNT are the most widely used to improve tear/wear resistance of the textiles. They can be mixed with many fibres precursor polymers such as polystyrene, polypropylene, or polyvinyl alcohol before spinning, or alternatively, applied to fabrics by spray or dip coating.

The Textiles with the NanoSphere® finish produced by Schoeller, presenting a naturally self cleaning effect; and the Tencel™ material based on nanofibrils of cellulose produced by Lenzing, which combines a good moisture management, reduced energy consumption, reduced growth of bacteria, antistatic behavior, and heat absorbing properties compared to common polyester making the material a good candidate for seat car covers. The nano-modified microfibre Evolon® by Freundenberg, used for headliner, dashboard, carpet backing, doors and the underbody shield, allows a weight and thickness reduction and better noise absorption. Elmarco's NanoSpider acoustic web offers similar sound reduction in addition to heat insulation and weight reduction.